Mysteries of Nature

EXPLAINED AND UNEXPLAINED

BY THE SAME AUTHOR

A ZOO IN YOUR ROOM
THE WONDERFUL WORLD OF MAMMALS: ADVENTURING WITH STAMPS
GOING TO THE ZOO WITH ROGER CARAS

Mysteries of Nature
Explained and Unexplained

BY ROGER CARAS

ILLUSTRATED WITH PHOTOGRAPHS

HARCOURT BRACE JOVANOVICH
NEW YORK AND LONDON

This book is for Pamela and Clay, who face the greatest mystery of all—life—with confidence and curiosity, the most essential ingredients.

Copyright © 1979 by Roger Caras

All rights reserved. No part of this publication may
be reproduced or transmitted in any form or by any means,
electronic or mechanical, including photocopy, recording,
or any information storage and retrieval system,
without permission in writing from the publisher.

Requests for permission to make copies of any part
of the work should be mailed to:
Permissions, Harcourt Brace Jovanovich, Inc.,
757 Third Avenue, New York, New York 10017

Printed in the United States of America

Library of Congress Cataloging in Publication Data

Caras, Roger A
    Mysteries of nature, explained and unexplained.
    Includes index.
    SUMMARY: Presents theories and explanations for
puzzling natural phenomena, including screams heard in
the nighttime forest and whales that strand themselves
on beaches and die.
    1. Zoology—Miscellanea—Juvenile literature.
2. Natural history—Miscellanea—Juvenile literature.
[1. Zoology—Miscellanea. 2. Natural history—Miscellanea] I. Title.
QL49.C26   596   78-20568
ISBN 0-15-256346-6

First edition

B C D E F G H I J K

HBJ

Contents

Illustrations   6

Introduction   7

1. The Saga of the Pacific Salmon   11
2. The Mystery of the Stranded Whales   17
3. Two Legends of Screams in the Nighttime Forest   20
4. The Mystery of the Elephant in Musth   24
5. The Mystery of the Lost Seals of the Antarctic   28
6. The Strange Case of the South African Sea Serpent   32
7. The Mystery of the Spitting Cobra   36
8. The Strangest Lizards of All   41
9. The Mystery of the Renegade Wolves   45
10. The Mystery of the Man-Eating Cats   50
11. The Mystery of Wadi Mukkateb   56

Epilogue: Where Does a Mystery Lead Us?   60

Index   62

Illustrations

Leopard (Holy Land Conservation Fund)   3
Mexican beaded lizard (Photograph by Roger Caras)   5, 41
Salmon (U.S. Department of the Interior, Fish and Wildlife Service. Photograph by George B. Kelez)   12
Beached whales (New England Aquarium/Cape Cod Museum)   16 (top)
Beached finback whales (New England Aquarium/Cape Cod Museum. Photograph by John Clarke)   16 (bottom)
Mountain lion (Photograph by Wilford L. Miller)   22
Elephant in musth (Photograph by Roger Caras)   25
Weddell seal (U.S. Navy. Photograph by Photographer First Class John Greenwood)   29
Sea serpent seen by officers (Drawing by Norman Arlott)   34
Spitting cobra   35
Gila monster (Photograph by Roger Caras)   42
Baying wolf (U.S. Department of the Interior, Fish and Wildlife Service)   45
Resting wolf (U.S. Department of the Interior, Fish and Wildlife Service)   46
Tiger   50
Lion (South African Tourist Corporation)   51
Lion drinking water (South African Tourist Corporation)   52
Markings on cliffs of the Wadi Mukkateb (Photograph by Roger Caras)   56, 57
Cliffs of the Wadi Mukkateb (Photograph by Roger Caras)   58

Introduction

When I was growing up in Methuen, a small Massachusetts town, I was surrounded by books about faraway mysterious lands. At night I dreamed of them. Their very names, China, Africa, Siam, Hong Kong, New Zealand, rang like chimes, calling me toward them. I tried running away a few times, though not very far, then resigned myself to waiting until after I had completed my education. At long last then, I began my journeys.

I have crossed the ocean sixty-eight times so far, and I intend to repeat the experience again and again because I can still hear the chimes of places I have not seen. I know of no music that is sweeter or more exciting.

While waiting for a chance to visit every corner of the Earth (something, of course, a person simply cannot do in one lifetime), I had a secret fear. I thought all the mysteries of the world would surely be solved and its secrets uncovered before I could explore it on my own. It was a foolish fear, as most are, for we are still centuries away from understanding the planet we live on. It holds more unrevealed secrets than there are grains of sand on the beach, many of which may not be discovered before the last human has faded into the foreverness of time and space. We may not be here on this planet long enough to learn all we want to know about it.

Still, these mysteries make life exciting. If you doubt that, consider how it would be if there were no place left to visit for the first time, no new person to meet, no new animal or plant or man-made marvel, like a space ship or submarine, to see, touch, and ask about. What if you knew it all already? It would be terrible! And boring. But as you know, there is no danger of its ever happening to you.

The French word *ambiance* describes the over-all feeling a place gives you when you visit it. I call this feeling texture, a combination of sights, sounds, and smells that makes each region like no other on this entire planet. That is what I look for when I travel, when I study a wild animal or search for a rare plant. It is what I seek on a beach in the Indian Ocean, what I hope to find on top of a mountain or deep in a cave, and what I am sure is beyond the next clearing in the jungle or forest.

A desert at night is quite different from your backyard or the nearest city park, and that yard and that park are different from a busy market street in Hong Kong or a wind-swept mountain in Hawaii. Certainly a jungle in Asia is full of surprises, but if you lie on your belly in the nearest plot of grass and spread that grass with your fingers, you will see that miniature marvels are there as well. Each place is unique and that is what you will eventually learn when you have traveled a great deal—that every locality offers many wonders. In this book we will examine some of these wonders. I call them mysteries of nature because scientists still don't really understand them. If your curiosity is aroused, you may try to solve one of these mysteries someday. And perhaps you will succeed!

Roger Caras
East Hampton, New York
August 1978

Mysteries of Nature

EXPLAINED AND UNEXPLAINED

The Saga of the Pacific Salmon

Salmon are known as anadromous fish. They live in the sea but migrate to fresh-water streams to reproduce. Fresh water and salt water make very different demands on animals because of the chemicals dissolved in them. Most fish must inhabit one or the other and cannot make the adjustments necessary to pass between them.

Salmon hatch out of their eggs in fresh water, rarely in a lake or river, almost always in a shallow stream over a rocky bottom. The female lays the eggs, and the male swims over them, spreading the milt. It fertilizes the eggs, enabling them to develop into infant salmon. When the young salmon hatch, egg sacs are attached to their bodies. They feed off the contents, staying down in the gravel at the bottom of the stream, where it is safe. After consuming the egg sacs, the salmon begin to feed on tiny water organisms. Then they start down toward the rivers and lakes that are nearer to the sea. Some salmon take a year to reach the open ocean, while others may take as long as two or even three years. It is a difficult and dangerous journey for the little fish. Larger fish lurk along the way, waiting to pounce upon and eat them by the millions.

Eventually some of the growing salmon do reach the sea and there begin their migratory journeys. It is this part of their lives that remains unclear, but many are found thousands of miles from the coast, deep in the sea, living

lives very different from those they led as young fish in sweet, fresh water.

Out of the thousands of eggs a female salmon lays, only two to four young are strong enough and lucky enough to survive all the dangers and hardships of growing up in streams, migrating to the oceans, and finally returning to their native gravel beds. Those few, though, are the best and toughest of their species, and they alone will produce the next generation. That is how nature works, of course, always picking the best to reproduce. That is how a species is kept strong in a very difficult and dangerous world.

There are salmon in both the Atlantic and the Pacific oceans. Atlantic salmon swim up streams in Europe and North America, reproduce, and return to the ocean. They repeat this cycle year after year. Pacific salmon do the same thing, with one enormous difference. They can do it only once. They all die almost immediately after spawning, as salmon reproduction is called. They go to the sea once, stay there from two to seven years, and return to their fresh-water streams once. Then it is all over.

We do know that the Pacific salmon are a newer species than the Atlantic variety. We know too that they descended one from the other as the older Atlantic salmon worked their way across the Arctic and moved down into the Pacific from the North. But why did the Pacific salmon change this way? Why must they die after one spawning while their ancestors, still living in the Atlantic, do not?

The reason may lie, according to one theory, in the kinds of coasts the different salmon face. There are higher mountains and hills in the West. Salmon returning from the Pacific Ocean must work much harder, in most cases, to reach a fresh-water stream with the right kind of gravel bottom. The only Pacific salmon that could make the trip and reproduce would be those that would work so hard to return to their native waters they would literally kill themselves doing it.

The flatter terrain in the East, around the edges of the

Atlantic Ocean, generally does not make such great demands. Atlantic salmon probably could not reproduce in Pacific salmon territory. They would not have the strength and energy to climb the rapidly running streams and leap up the endless waterfalls even once as their Pacific descendants do.

This, however, is still a theory, and although it is probably the best explanation at the moment, it is a long way from being accepted as scientific fact. Scientific knowledge often starts as theory, which can develop from an intelligent, or sometimes just a lucky, guess.

There is another mystery about both Atlantic and Pacific salmon, and many theories attempt to explain it, but none can be accepted yet. This mystery concerns the fish's navigation, or migration, as it is properly called. Salmon return, as far as we know, to their own gravel beds, the places where they themselves hatched. How does an infant salmon leave its place of birth, migrate hundreds of miles to the sea, travel in the ocean for several years, and then find its own gravel bed again so that it can reproduce?

A salmon returning from the deep sea moves along the coast and probably passes the mouths of scores of rivers. It must first pick the right river. Then it must start the difficult trip upstream. On a long and large river it may pass dozens of tributaries, or smaller rivers that feed the larger one. It must pick the right tributary. Then a tougher climb begins. There may be waterfalls and rapids, but the salmon push on toward their native gravel beds.

How do they do it? Some people say they navigate by the sun, others say by the moon and stars, while others believe they have extremely fine taste and can literally taste their way home! According to this theory, as salmon move downstream and out to sea they record all the tastes along the way like computers and, again like computers, they work it out backwards as they return. It is true that no two streams in the world taste the same. Different amounts of chemicals dissolved from rocks and soil would make each unique. It would have its own chemical autograph. But again, that is theory. There are two reasons to doubt it: a salmon would probably be unable to taste its way home at sea; and since man is constantly changing the chemical ingredients in rivers and streams, their taste would be constantly changing too. Of course, salmon could use two or more methods of navigation during the homeward migration. They could use one system in the sea, one in the large rivers, and then finally taste their way home in the streams. All of that is possible.

Ships equipped with computers navigate automatically. The computers figure out where to go by "remembering" where they have been. This is particularly valuable in nuclear submarines that stay down deep for long periods of time. Could salmon have a similar kind of navigational system?

Other forces that may help salmon navigate are gravity and the Earth's magnetic field, both of which play a role in the behavior of some animals.

Future studies may shed more light on this and many other mysteries of the sea, which in turn may help us to better understand ourselves. Every living thing has evolved from the primeval cells that were sparked to life in the waters that once covered the earth, and the same planetary forces affect us all.

The Mystery of the Stranded Whales

2

Aside from man, whales may be the most intelligent animals on Earth, although some people believe great apes are brighter. Is there proof of whales' intelligence?

Man has held these mammals (they are not related to fish) in aquariums for years—not the great whales, of course, because of their incredible size, but the smaller ones. We call them dolphins, porpoises, killer whales, belugas, but they are all in the whale family.

We have trained many captive whales to do tricks and follow instructions; and we have tried very hard to understand their communication system—an elaborate vocabulary of sounds, if not words. We have been only partially successful. If you stop to think about that, you will realize that it is amazing, because whales do comprehend the signals we teach them. If they did not, how could we train them? It is interesting, I think, that men have worked with these sea creatures in aquariums for decades, and whales are way ahead of us in our mutual efforts to understand each other. We are only now beginning to interpret some of their sounds and sonar signals. This must tell us something.

Whales, whether they are the smaller porpoises and dolphins or giants like the sperm and great blues, live in communities. We know they care about one another because when one is in trouble the others in the group try to help. Their methods of communicating are complex, as we have suggested, which would come about only among animals that relied on one another. The more interdependent animals are, the more complicated their communications.

Whales travel enormous distances, often thousands of miles. Deep in the ocean they call to each other and, we think, even sing. They all sing a similar song for a while, vanish into the distant oceans, and return a year later, singing a different song. In addition, they use sonar signals, sending out sound waves, which hit nearby objects, then bounce back. That is how they sense the sizes and shapes of things in the waters surrounding them, such as food, plants, ships, enemies, or anything else in their vicinity.

Descendants of land animals, whales require air just as you and I do. They can stay underwater a very long time, possibly up to two hours in some cases, but eventually they must surface to breathe. Because their bodies are supported by water, they have grown to sizes that would be impossible to maneuver on land. A bull blue whale, for instance, can weigh 300,000 pounds and be well over one hundred feet long.

Also, its lungs are enormous. In order to fill them the whale must expand its rib cage just as you do. On land the whale would have to push its incredible weight up for every breath. After a very short time it would become exhausted and perish. Its own weight would suffocate it. In addition, a whale must not be exposed to sunlight for very long. It gets blistered and burned. A whale is supposed to be wet all over all of the time.

Thus for a variety of reasons whales have become prisoners of their environment. They can never leave the water again, unless, of course, after millions of years they somehow evolved back into land animals. That might happen only if the seas should become overcrowded or inhospitable.

One would think, then, that whales would avoid shallow water and beaches. Yet every year scientists report that hundreds of whales are killed in "strandings." In some cases only one great whale may strand itself. Other times, an entire herd, or pod, as a group of whales is often called, numbering from ten to thirty animals, enters shallow water and pushes up onto a beach. In an effort to help, men and women have tied ropes to the animals' tails and pulled them back into the water with boats. The whales, however, alone or in pods, insist on driving themselves back into shallow water to their death on the beaches. It is a tragic sight. Records of whale strandings have been kept for scores of years, but we still do not understand why they occur. It was once thought that the smaller whales were chased into shallow water by a great hunting whale (actually the largest of the real porpoises), often called the killer whale. But when accumulated records were examined, it was found that killer whales strand themselves just as often.

It has also been suggested that whales will follow a leader even to their doom if the leader goes off course.

The majority of marine biologists seem to think that stranded whales are sick, victims of a virus or parasite, which has gotten into their ears and made their usually superb navigation system malfunction. We don't know whether that is true either, any more than we know how the animals actually navigate.

A lot of theories, ideas, and suggestions have been offered, but the stranding of whales remains a mystery. It may be a long, long time before we really understand it.

3 Two Legends of Screams in the Nighttime Forest

Since the beginning of time man has invested both the forest and the night with mystery. We humans cannot see well at night so the animals that come out then are those with which we have the least experience. There is another factor. When one of our senses is impaired the others become much sharper. If you wear a blindfold for a couple of hours your hearing will become much more acute. So it is at night. We hear more because we see less, and what we hear often gives rise to flights of imagination.

What the night is to time, the forest is to place. Forests and jungles are harder to see in than deserts or plains, therefore we assume stranger things go on there. When one combines nighttime *and* the forest, the most incredible things can come into being.

As the early European settlers in North America began moving west as well as north and south, they soon began to encounter a strange and terrifying sound. It occurred in the forest they said, and very often at night. It was a scream so terrible that brave men crouched by their fires and spoke of the devil. Some said it was like the cry of a woman being killed, combined with the scream of a steam locomotive. That description came after the steam train was invented, but stories of the nighttime sound of the American forest go back further than that.

Often woodsmen attributed the sound to the Indians. They said it was the sound of magic men, who could transform themselves into hideous monsters, then run or perhaps fly through the forest, killing enemies with nothing more than their horrible voices.

For over a century the sound was described again and again, and people who never went into the forest at night laughed. "Old woodsmen's tales," they said. "Pure nonsense . . ." "Drunken old trappers with nonsensical tales," they proclaimed. "They only want to sound braver than they are."

Then at last the mystery was solved, although no one remembers who first figured it out. It was the female mountain lion calling to her mate.

A mountain lion (it is known by many names—puma, panther, painter, cougar, and others) usually has a small voice. Even a full-grown male normally does little more than chirp like a canary. He can growl and snarl and spit and hiss, but usually he just chirps.

When the female wants to breed, however, she announces it to the world. Since a mountain lion is a solitary animal, the female's voice must carry far. It sounds like a fifty-pound piece of chalk being scraped on a blackboard, along with a truck grinding its gears, plus escaping steam. Actually there is no way to describe it except terrible. Evidently male mountain lions get the message. If they didn't there wouldn't be any baby mountain lions.

There are legends of such sounds in other countries as well. In Sri Lanka (then known as Ceylon), people fear one sound as much today as they did a thousand years ago. It is known as the call of the devil bird.

On that beautiful island in the Indian Ocean there are many thick forests that are rich in wildlife. Elephants and water buffalo are common on the trails. There are sambar

stag and spotted deer, leopard, many, many kinds of spectacular birds, wild boar, and other smaller animals. At night some animals that hide during the day begin foraging for food. The loris, with its huge eyes, and the very cranky sloth bear begin their movements from place to place. There are crocodiles in almost every pond in the jungle clearings.

Somewhere in the jungle there lives a creature whose night scream has inspired books of legends. It is called half devil and half bird. A man was lost in the forest, according to one legend, and he heard the scream of the devil bird. Although he had been a young man when he entered the forest the night before, his hair turned white, his face became lined like a very old man's, and he was never able to speak again. Many people who have heard the scream are said to have gone mad or have even died of heart attacks. It is commonly said that if one hears the scream of the devil bird, he will never be the same again.

What *is* the devil bird? That, too, was finally solved. It is nothing more than the Ceylonese eagle-owl, a large forest-dwelling owl that does indeed have a harsh, unpleasant voice. Its voice, combined with the known presence of sloth bears, wild elephants, water buffalo, leopards, pythons, cobras, and the darkness of the jungle might indeed startle a person.

The big difference between Asia and North America is that here, once the identity of the screaming puma was known, stories about the night-forest devil-monster died away. In Asia they persist, although educated people have known for a long time that the source was nothing more than the call of an owl.

4 *The Mystery of the Elephant in Musth*

Elephants are generally peaceful, social animals. They rarely fight seriously among themselves or attack other animals, including man, unless they are threatened. Occasionally, though, a mature elephant that has led a quiet, tranquil life suddenly goes mad, destroying everything in sight. They call such an elephant a rogue.

In Asia, I was told that such rampages periodically affect the same elephant. They call this periodic insanity musth. When an elephant is "in musth," everyone must stay clear. Natives believe the animal should be left alone until it recovers.

Once, on the island of Sri Lanka, I visited a sawmill two hours after a working tusker went mad and leveled the entire mill. The fences were knocked down and the buildings flattened. There was ruin everywhere. Since a good working elephant in Sri Lanka is worth $5,000—more than it costs to rebuild a sawmill—there was no attempt to hurt the elephant. An old man who could handle dangerous animals took it off into the forest, where he hoped to calm the beast. The man was known to have powers over animals and was called whenever there was trouble with an elephant or a snake. He would talk to it, and everyone believed it would obey. Natives called him a magic man. The bull, they said, was in musth and was best left alone by anyone but the magician. I was advised against following them into the forest to take photographs because the bull would charge if it smelled or heard my approach. (Elephants don't see well and don't rely on their eyes.)

One of the first signs that an elephant is approaching musth is a discharge from glands on its temples. There is one on each side, with a small, usually invisible, opening in the skin leading to it. Dark stains run down the animal's cheeks from those two openings. When that happens elephant workers expect trouble.

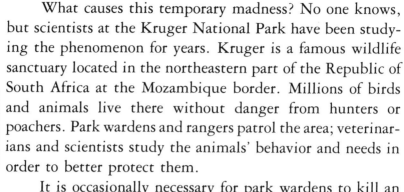

What causes this temporary madness? No one knows, but scientists at the Kruger National Park have been studying the phenomenon for years. Kruger is a famous wildlife sanctuary located in the northeastern part of the Republic of South Africa at the Mozambique border. Millions of birds and animals live there without danger from hunters or poachers. Park wardens and rangers patrol the area; veterinarians and scientists study the animals' behavior and needs in order to better protect them.

It is occasionally necessary for park wardens to kill an elephant for any one of several reasons. It may be old and crotchety or young and especially aggressive. In either case the animal might be dangerous to visitors. Wardens do not wait long to eliminate an elephant that has attacked buses and cars on the roads. Seriously wounded or sick elephants drift into the park from unprotected areas across the border, and they must be killed so that they won't suffer. Often they have been the victims of poaching attempts. Also, elephants are found dead from old age, disease, or, rarely, from a fight or accident. One way or another in a place as large as Kruger National Park—nearly 5 million acres—there is no shortage of dead elephants to study.

An area near park headquarters is a fenced-off compound the visiting public is not allowed to enter. It is a scientific research center where sick animals are taken for protection and where plant life and the animals that live off it are studied.

In the corner of one room in the veterinary section on a gray steel file cabinet are dozens of little bottles, each containing a single thorn or small twig. A label on each bottle tells scientists when and under what circumstances the specimen was collected. Those seemingly unimportant bits of vegetable matter and their connection with elephants form the basis of one of the most confusing mysteries I have ever encountered in nature.

The veterinarians had been examining dead elephants in the park for years, and they paid special attention to the musth glands. They had discovered a most amazing thing. Whenever an animal appeared to be in musth, whenever its cheeks were stained, there were thorns or twigs inside the musth glands.

The musth gland has a peculiar shape. Its tiny opening from the outside leads to a twisting narrow canal, then to the small gland deep under the skin. If you were to try to push a flexible wire in through that opening and canal, you would have a hard time.

Scientists believe it is impossible for thorns and twigs to get into those canals and glands by accident. Elephants rub up against and push through many trees and thorny shrubs while feeding, but the veterinarians are convinced that the twigs and thorns are placed inside the musth glands by the elephants themselves.

Clearly, no other living creature could insert those bits of woody material into the animals' temple glands! As dexterous as an elephant is with its trunk, the only appendage it could possibly use, it hardly seems possible for it to pick a single thorn off a bush, find that tiny opening (which it cannot see), and work it in through the canal. That, though, strange as it may seem, is what must happen.

Then there is a bigger question. Why would an elephant do that? When you ask that question, the Kruger veterinarians look at each other and shrug. No one has even a clue.

5 *The Mystery of the Lost Seals of the Antarctic*

The Antarctic may be the least hospitable place on Earth. People confuse the two ends of the Earth, so it might be worthwhile to distinguish between them. In the North is the Arctic, a great frigid ocean basin about 10,000 feet deep. Somewhere in that ocean is the North Pole. It isn't really a pole, just a place.

The *Ant*arctic is in the South. It is a huge continent of almost 6 million square miles, with mountains and valleys and flatlands. It is covered with snow and ice, and somewhere, high upon what is called the polar plateau, is the South Pole. Again, that isn't a thing, but a location. No one is certain that these magnetic poles have ever been accurately found. They can be located only by making extremely fine measurements, and we are not sure our instruments are precise enough yet. We have certainly come close, though. Probably within a few feet, or even inches.

There are other differences between the two areas. People live in the Arctic, while the Antarctic has never been inhabited by man, though it has been visited for long periods by explorers and scientists. There are large animals in the North, as well. Polar bears, grizzly bears, moose, caribou, and reindeer are found in the Arctic, along with foxes, wolves, and many kinds of birds. The largest land animal in the Antarctic's vast area is a wingless mosquito! The only other land animals are a few even more primitive creatures without backbones. Once, when the Antarctic was

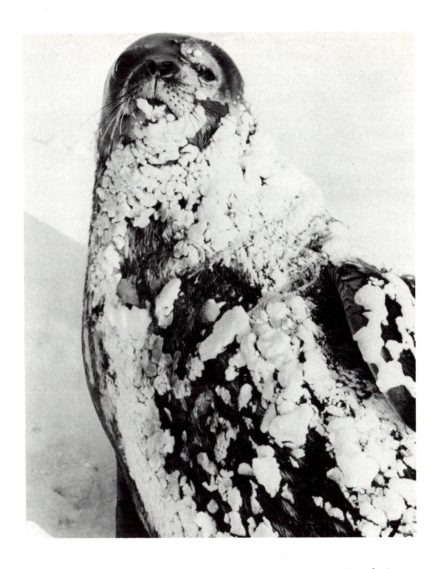

warmer than it is now, prehistoric land animals lived there, but not since nature turned the continent into Earth's deep freeze.

There are sea animals, though, and they abound. The waters are inhabited by billions of small red shrimp called krill, and the greatest whales in the world go to feast on them. Killer whales hunt other sea mammals. Other marine animals include several kinds of seals and penguins from the Southern Hemisphere, along with a few species that are found only around the icy edges of the Antarctic continent. There is also a gull-like bird known as the skua and a tern that migrates between the Arctic and the Antarctic every year, making a trip of close to 20,000 miles. That is all.

The seals and penguins do come out of the water. The penguins nest on the ice, and the seals emerge from air holes and lie around for hours at a time. It is one of these seals, the Weddell, that behaves very strangely.

The Antarctic starts at sea level, where the sea itself is often frozen to a depth of eighteen to twenty feet. From that frozen sea shelf the continent climbs upward toward giant mountain ranges and a vast polar plateau. At the South Pole the ice and snow is two miles deep. The wind screams off that plateau and roars across the frozen ice shelf below. Winds blow harder there perhaps than anywhere else on Earth, up to 200 miles an hour—like two hurricanes rolled into one. The coldest temperatures ever recorded were measured on that plateau. They have been known to plunge past 127 degrees Fahrenheit below zero. Temperatures that low with 200-mile-hour winds create a wind-chill factor that can kill an exposed human being in seconds. Fortunately, when I visited the South Pole, a heat wave was in progress. It had warmed up to seventy below zero.

Nothing grows on the slopes and glaciers leading to the plateau. The valleys are deep and rugged. The ice, which forms glaciers and flows down toward the sea in an endless stream, is fractured and cracked into crevasses a mile or more deep. There is nothing to eat, certainly nothing for an animal that normally depends on the sea for its food.

Here is the mystery: time and again Weddell seals (and occasionally other seals, too) have been found dead from starvation and exposure miles from the sea. Their movements have been tracked from the safety of the food-rich ocean to a senseless, suicidal journey inland to the slopes and up toward the South Pole. There was nothing for them along the way and nothing for them even if they reached the pole. As far as we know, none ever has.

What could attract a seal, drawing it away from food and safety in the rich pastures of the sea to certain death in that cold lonely world? We do not know. Some people say the seals get lost and confused and simply wander in the wrong direction. That doesn't seem likely, since seals navigate over great distances at sea. They would soon discover something was wrong.

Some believe these animals are diseased, which causes them to act strangely or to lose their way.

Whatever reason lies behind this strange occurrence, every year a number of seals (no one knows how many, for the area is vast, and bodies are quickly covered by snow and ice) crawl awkwardly out of the sea and flop along on the ice. They may flop and wriggle for miles, eventually reaching the edge of the continent itself. They move along the edge of the fractured ice until they find an opening they can force their fat bodies through, and then they begin their journey to death.

I know from experience that when you come upon such a seal you cannot help but feel sorry for it. Life has played a cruel joke on a helpless animal that is lost and far from the safety it has known in the sea since its birth.

6 *The Strange Case of the South African Sea Serpent*

For as long as man has been going to sea there have been stories about mysterious and, often, dangerous animals that were unknown to science. Many of these stories have been exaggerated by time and retelling, a small incident becoming a big one, a small animal becoming a monster.

These wild and woolly stories were told for many reasons. Life at sea can be dull, and men on sailing vessels were often at sea for months and even years. They made up stories to pass the time and to keep their minds active. Since few sailors in the old days were able to read, storytelling was important.

Then there was the homecoming. If you had been sailing around the world for several years and arrived home to a big reception by your family and friends, could you say "Nothing" when they asked you what happened during your trip? A sailor was almost obligated to tell a story or two about his adventures.

Finally, there was the possibility of mistake. When the sea is calm, as when it is rough, things can be difficult to identify. A line of porpoises or group of low-flying sea birds can appear to be something else, perhaps even a sea monster or sea serpent, as this "thing" has often been called.

For centuries stories about things that probably were not there most of the time came to have a kind of reality. People said they saw them, people wrote that down, and as other people copied the stories over and over again, the creatures became larger and larger. People all over the world believed that a huge monster that science had never identified (because scientists had never gotten a close look at one) was alive in the sea and causing trouble for sailors. Many would think you were quite mad if you questioned the serpent's existence. He surely existed, in their minds.

Most sea-serpent stories could be explained away. Those who told them had heard the stories at fourth and fifth hand, but they hadn't seen the thing themselves; they were known to be storytellers; or perhaps they were a little drunk when they claimed they had seen a sea monster. Most of the stories could be put aside for those reasons—but not all.

During the 1880s a number of sea-serpent "sightings" were reported around the Cape of Good Hope in southern Africa. The serpents fitted the general descriptions found throughout all of man's seagoing history, and there was no reason to take them seriously. They were amusing tales with perhaps a few cases of mistaken identity and a few cases of downright storytelling. For me, at least, there was no special reason to believe them until I came upon an old and badly worn book.

A World of Wonders was edited by a man named Albany Pyntz and published in London in 1845. Someone who had owned the book long ago had used it as a scrapbook. Inside the covers and jammed between the pages was a strange collection of newspaper clippings and notes about mysterious happenings in the world. Whoever made the collection had more than a passing interest in the subject matter of the book.

Among the clippings was a real treasure, in two parts. One was an original drawing of a sea serpent, which had never been printed anywhere else. The second part was a letter describing the monster. Both were originals, and the drawing had been made with faint-colored pencils. I was so fascinated by these documents that I began checking the records. Nowhere could I find any other information about an occurrence on August 12, 1881.

That day several high-ranking British military officers boarded a steam launch near Cape Town, South Africa, and set out to inspect the shore batteries that protected the great bay from enemy attack. The Governor General of South Africa, General Sir Leicester Smyth, was on board, along with several other officers, who were later to become admirals and generals in World War I. A more respectable group of men could not be found. They were not drunk, ignorant seamen. These were the best men the British army and navy had to offer.

Near a small island named Roman Rock they saw something that apparently surprised them. They approached this "thing" in the water at slow speed, and although it began to move away, they got a good look at it. Since they were inspecting shore positions, we can assume they had telescopes with them. Whatever it was, they saw it clearly

enough to have a drawing of a large sea serpent made.

Although the incident was not made public, a few months before and afterward, other sightings of so-called sea serpents were reported along the coasts of South Africa. They could have been discounted were it not for this one sighting, because it was made by a team of clear-minded, high-ranking officers from a nearby boat on a clear day. The officers were neither drunk nor stupefied from months of sea duty.

I have visited the Roman Rock area and looked at the very same shore artillery batteries they inspected. I have checked newspapers of the area and have found no reports of that sighting. It was apparently kept a secret until the drawing and letter were found in that strange old book on nature's wonders.

We cannot be certain of what did happen. We cannot state that an unknown sea animal of considerable size exists just because those papers appeared after so many years. But because of the notes and the caliber of the men who reported their observations, we must at least consider the existence of a sea serpent. A thousand men may be mistaken about something, and another 10,000 men may lie about it. It doesn't follow, however, that one man or a group of men are not telling the truth. We still don't know the truth about sea serpents, but how exciting it will be for the man or woman who makes the final discovery—if it is ever made.

A very similar and more familiar mystery surrounds another creature, known as the Loch Ness monster. It is said to appear in one or more lakes, or lochs, in Scotland, but also in Canada, Sweden, and Russia. Many people are investigating that mystery. Perhaps its solution will help us to understand the sea serpent as well.

7 *The Mystery of the Spitting Cobra*

In all the world we have discovered 2,700 different species of snakes. Like lizards, they are found on every continent except the Antarctic.

Everyone, of course, immediately thinks of deadly venom, or poison, when the word "snake" is mentioned. Only 415 species, or about 15 percent, that we know of are really venomous. That isn't many, not enough to justify killing every snake in sight and fearing the others. Even most of the venomous ones offer no problem to man.

Of those that are truly venomous, none conjure up stranger pictures than cobras. The name cobra seems to imply mysterious and dangerous places. While cobras do come from faraway places—Africa and Asia—much of the danger has been greatly exaggerated.

Cobras do not spread their *heads* to form that famous hood. The cobra's head is small and bent over in front of the hood. The snake spreads the ribs in its neck, behind its head. Its purpose is simple. A cobra is an ordinary-looking snake in its normal, relaxed position, but when it hoods, it looks much larger, much more threatening to animals (including man) that might harm it. Like all venomous snakes, it tries to avoid trouble. The very least that can happen to it if it fights a larger animal is that it will have to waste some of its precious venom, which the cobra would rather use to catch its food.

When we say "rather" we don't mean that a cobra

thinks about things and makes a choice. We mean it instinctively does one thing rather than another. It knows without thinking what is good for it and what isn't.

A cobra's fangs are short, much shorter than a rattlesnake's, for example. Only about five out of every hundred people who are bitten by a cobra die. Tales that certain death follows every bite are just that: tales. This is not true of an Asian animal called the King Cobra, for that is a different snake altogether. It is the largest venomous snake in the world—almost nineteen feet long—and it does kill a higher percentage of the people it bites. Not so the other, smaller, cobras, however. And those are the snakes we are discussing here.

There are between fifteen and eighteen kinds of cobras in Africa and Asia. Although they are deaf, they can sense the movement of heavy animals nearby through vibrations in the ground. If they have a chance, they get out of the way. If startled or approached suddenly, they rear up, spread their hoods, and try to look as dreadful as possible. If the intruder backs away or stands very still, the cobra lowers itself to the ground and tries to crawl away. It is interested in rats and mice, not elephants and human beings. It wants to bite only something it can swallow whole. It cannot chew.

A few cobras have special abilities, though. They are called spitting cobras. One is found in Malaya, a long peninsula in eastern Asia (this snake has the lovely scientific name *sputatrix*), one is found in Thailand, and a couple are found in Africa. As far as we know, no other snakes in the world can spit their venom.

How is it done? A snake's fang is a hollow tube like a hypodermic needle. There is a hole at the top, under the snake's gum. Venom flows into that hole, down through the fang, into an oval opening at the bottom. If a snake wants to inject venom when it bites (sometimes a snake just gives a warning "nip"), it tightens the muscles behind its head, pushing the venom into the fang and out through the bottom.

The spitting cobra has a different system. Instead of a long slender opening at the bottom of the fang, like a doctor's needle, it has a small round hole part way up from the tip. Inside, the channel along which the venom flows takes a sharp turn toward the hole in the front edge of the fang. When the venom squirts down the fang, instead of going out the bottom it hits that shoulder inside and squirts out rapidly. A mist is formed as the venom droplets break up in the air.

The spitting cobra always aims for the eyes of its target. It is not as powerful or accurate as is often believed, but from a distance of six or seven feet a large spitting cobra can hit a man in the eyes. The venom burns, blinding the victim temporarily and giving the snake a chance to get away, for the spitting cobra uses the venom it spits to protect itself, not to hunt for food. If the venom is washed out of the eyes very quickly and the proper ointment is used, the blindness passes.

Snakes are vulnerable animals. Birds, mammals, and other reptiles hunt them for food. Man hunts them for their skins, for food in some parts of the world, and for zoos and circuses. It is not easy for a snake to survive in the wild even if it is a venomous snake. It doesn't do the snake any good to bite a hunter if the man crushes it before he dies.

To solve the problem of being bumped into accidentally by large and dangerous animals, rattlesnakes have de-

veloped their noisy, warning tails. Some harmless snakes have learned to make cobralike hoods to fool enemies. Others waggle their tails nervously in dried leaves, making sounds like those of rattlesnakes. But in all the world of venomous snakes only members of one out of every hundred species can send their venom out like protective jet-powered weapons. All other snakes must be close enough to bite in order to use their venom. If they are close enough to bite, they are close enough to be bitten, stepped on, gored, thumped, kicked, picked up, and tossed. Clearly, a snake is better off if none of these things are done to it.

Most species of animals have good defenses. Flying is a

good idea for a bird and a fine way to escape from enemies. With very few exceptions, all birds fly. Being able to imitate its background is a good way for an animal to escape enemies, and tens of thousands of animals are equipped with protective coloration. Being able to run when attacked is useful, so all hoofed animals are fleet. Being stealthy and able to stand perfectly still for long periods is helpful to hunters, and all cats have those abilities. Nature, in short, passes a good idea around.

Why, then, has venom-spitting been such a jealously guarded skill? If it works so well for a few species of cobra (and it does), why can't the others master the skill? It seems as if most, if not all, venomous land snakes could benefit from the ability to keep their enemies at a safe distance. That is not the way nature has worked things out, however.

The Strangest Lizards of All

There are about 3,000 kinds of lizards in the world. Some are very small, a little more than a couple of inches long. And some, like the Komodo dragon, are monsters capable of killing a pig or a deer.

Some lizards are vegetarians; they eat only plant matter. Others eat insects. Some of the most interesting lizards, though, eat meat.

Although lizards are reptiles, they are far more than "snakes with legs," as some people seem to think. They have ear openings (no snake has ears) and four legs in most cases. There are a few legless lizards, animals we call glass snakes, for example, though they should be called glass lizards.

Wherever lizards are found—in every continent except the Antarctic—there are stories about poisonous lizards, or, more properly, venomous ones. These lizards are said to be able to sicken or kill if they bite. Some people are afraid to touch one or even look at it. They take such tales seriously. Once, when I was visiting the island of Sri Lanka in the Indian Ocean, I reached out to capture a harmless little lizard sitting on a rock. I was curious about it and attracted by its bright-blue color. I wanted to examine and then let it go, but I never got a chance to do so. My secretary, a fine, strapping fellow named Thambi Amirthamayagam, charged into me like an express train, knocking me off a rock and down a steep slope. My clothing was torn, and I was cut and bruised. I was furious, of course, but I quickly realized that

Thambi was frightened. He, too, had tumbled down the slope and over the jagged rocks after me. I was lucky to roll out of his way before he landed on me, which, I am sure, would have done even more damage.

"Whatever made you do a fool thing like that?" I asked.

"The lizard, sir. If you had touched that fellow you would have died very soon."

Thambi wasn't joking. He believed it. Although he is an educated man who can speak and write four languages, he believes, as so many people do, that the world is crawling with deadly lizards that are waiting to kill creatures as large as man.

Are there any venomous lizards?

Just two, and both come from North America. One is found in the United States, the other in Mexico. Of our 3,000 species of lizard, those are the only two that live up to their fearsome reputations.

The two poisonous lizards belong to a family known to scientists as the Helodermatidae (pronounced "hell-o-derm-at-i-dee"). The lizards are known as the Gila ("hee-lah") monster and the Mexican beaded lizard. They look very much alike. They are large animals covered with black and pinkish beads. They have blunt heads and powerful jaws. The venom they produce in their mouths flows forward along grooves when they bite down. It is this powerful venom (it is really like the saliva in your own mouth) that enters the wounds their sharp teeth make.

At least eight people we know of have been killed by this venom, a large number, since both beaded lizards and Gila monsters are rare and shy. They don't approach people in the wild, and everyone who was ever bitten (about forty people) was handling one of the animals at the time of the accident. The lizards have bulldoglike jaws. When they bite, they hang on and the venom runs into the wound. It may be ten minutes before one lets go. The injury is very painful and very dangerous. Obviously, Gila monsters and Mexican beaded lizards are not meant to be kept as pets. In the wild they are protected; it is unlawful to disturb them. They are becoming rare because there are now roads through the desert wildernesses in Arizona and Mexico, where they live. The animals can't seem to learn about the danger from cars, and some are killed on the highway every year.

Here is the mystery. Both of these lizards eat what we call pink mice—mice so young they haven't even gotten their fur yet—and fledgling or newly hatched wild birds. The lizards clomp along in the desert, poking under rocks and the few bushes that exist there, looking for mice and bird nests. When they find them, they rob them of their young. It isn't a pretty picture, but the lizards must eat.

The job assigned to them by nature is to keep populations of these other animals under control. The lizards' way of life is nature's scheme, not the animals'.

The jaws of both lizards are so strong they hardly need their venom at all, as snakes do when they capture their prey. The prey of the Gila monster and Mexican beaded lizard cannot run away, for they are still too young.

All other animals with venomous *bites* (about 400 kinds of snakes and a few very small mammals known as shrews) are equipped with venom for food-getting first and for self-defense second. So why have these lizards been given venom powerful enough to kill grown men? Some people say the venom enables the animals to defend themselves against man, but that can't be so, because both Gila monsters and beaded lizards were here long before the first man ever walked upon this planet.

We like to think nature doesn't waste energy, that the abilities animals and plants evolve serve a need. What is the need for the effect of Gila-monster and Mexican-beaded-lizard venom? No one knows, but we have a theory.

Perhaps, long ago, larger animals hunted these lizards. Perhaps they needed that venom to protect themselves against these now unknown predators. If this theory is correct, whatever that animal was, it is almost certainly extinct. The venom of the two poisonous lizards, then, would enable them to live in a world that no longer exists.

The Mystery of the Renegade Wolves

9

Most mysteries in nature are not tragic. They are just interesting. There is one, though, that has had disastrous results. It is the mystery of renegade wolves.

To understand the extent of the mystery, you must know the character of the wolf. A highly social animal, it is unhappy when living alone. Both the male and the female crave the company of others of their kind and generally live in peace with one another. Sometimes there is only a single pair of wolves and their cubs; in other cases fifteen or twenty travel and hunt together.

Nature has built a marvelous device into the nature of wolves that keeps them from harming each other. They have enormous teeth and extremely powerful jaws. They are quick and smart. If uncontrolled and in competition for mates or food, they would soon kill each other off. A dog fight is bad enough; a wolf fight must be terrible. But wolves very rarely fight. They do challenge each other, though. Then an outsider may think they are tearing each other apart, but no such thing is happening. When a wolf is "dominant," when he is king of the pack, he is known as an Alpha Wolf. The second male is the Beta wolf. (Alpha and Beta are A and B in the Greek alphabet.)

If a lesser wolf in the pack, from Beta on down, challenges the Alpha wolf, the great male will throw himself at the challenger, making the most ferocious sounds imaginable, accompanied by terrifying facial expressions. The ears

are laid back, the lips are curled, the teeth are showing. It is almost as if fire were coming out of the Alpha wolf's eyes. He pushes his teeth into the fur of the other wolf's throat. They are poised right over the carotid artery, the major blood vessel carrying blood to the brain. If the Alpha wolf bit down and hit that artery, the other animal would be dead in minutes.

The amazing thing is that the Alpha wolf not only does not bite down, he *can't*. An invisible power holds him back. At this point in the challenge display, the lesser animal has turned his head over on its side, dropped down onto his front elbows, and made his artery easily available to the dominant animal. He is also whining and moving in such a

way as to announce, "I surrender." As long as he does that the Alpha wolf is frozen in place by his inherited behavior patterns. He can snarl, growl, look and sound horrible, but he cannot and will not bite. In normal wolf society that is how most disputes are settled. Actual fighting is rare and usually occurs when strange wolves try to invade the hunting territory of an established pack.

Other things about wolves make them interesting and special as well. They are superb parents. Adults mate for life and share the task of raising the young. When the female is in the den with cubs, the male goes hunting alone or with other members of the pack. He returns with chunks of meat in his stomach, but he can hold back digestion. He regurgitates the meat, fresh and ready to be eaten, for his mate and their cubs. Later, when the male and female start hunting together again, they may leave the cubs alone or with another member of the pack, a member lower in the social order. Both parents bring meat home.

Wolves seldom kill more than they need. They are not the superb stalking killers the leopard and the tiger are, nor are they as swift and sure as the wild cats. They work hard, and they work together. Very often the prey sought by wolves is large and dangerous, notably moose. Wolves try to find a sickly moose or any animal that has been slowed down by age, and they kill it for food. They don't take unnecessary risks or waste energy killing animals they don't need to eat. That is the way they have evolved. It would be bad for wolf survival if they wasted meat needed the following week or month on a killing rampage, so nature has made them behave otherwise.

Every now and then things go wrong, even in nature.

That has happened to wolves from time to time, and that is how stories about them started. Periodically, a single wolf goes on a rampage and kills wastefully, wantonly, and without purpose or good. Such wolves are renegades and usually live and hunt alone.

There have been a number of these wolves through history. In this century a small white wolf became known as the Custer Wolf because it spent its most destructive years in Custer County, South Dakota. It was finally killed in 1922, but it took government agents five years to track it down. It is said to have killed or mutilated a great many horses, cows, and sheep.

In Revolutionary days a renegade in Connecticut came to be known as Old Put's Wolf. It had the bad luck to get into the sheepfold of a Revolutionary War hero named General Israel Putnam. The general and his men tracked the wolf to its underground den, and then the military leader crawled into the cave and shot the wolf in the head with his pistol. Between Old Put's Wolf and the Custer Wolf there were others that became legendary: Old Doc and Three Toes, for example. They were so unusual in their behavior that they were given names we recall to this day although some have been dead for almost 200 years. If nothing else convinces us that random killing by wolves is unusual, the fact that the renegades are so well remembered should do it.

Unfortunately it is this strange, really insane, behavior of a *few* that has given all wolves an evil reputation. Stop to think about our own language and figures of speech. If there is poverty, we say "the wolf is at the door." If a man is unfaithful to his wife or girl friend, we call him "a wolf." A wolf huffs and puffs and blows down the house of the poor pig. Little Red Riding Hood has trouble with a sly, deceitful wolf, which could be referred to as a "wolf in sheep's clothing."

In cartoons the wolf looks like an incredible monster waiting to trick and devour everything that lives. But the facts are different. The wolf is a large, splendid-looking wild dog, trying to earn a living for itself, its mate, and its cubs. It doesn't usually fight, although it defends its cubs to the death. It is shy, and it takes from the land only what is essential to its survival.

Why is the misunderstanding caused by the occasional renegade tragic? Because mankind has been misled by these fables. We believed our own nonsense, and that is how we justified killing most of the wolves that once lived on our broad, beautiful continent. We fear wolves as if we were children who believe in cartoon characters on television. In fact, there are no records in all of North American history of wolf attacks on human beings. They are legends, just like "The Three Little Pigs" and "Little Red Riding Hood."

10 *The Mystery of the Man-Eating Cats*

No animal "characters" in adventure stories, hunters' tales, or accounts of exploration have thrilled more readers than the so-called man-eating cats. Terrifying tales of great cats that stalk man in the night and carry him off amid screams of pain and despair have come from Africa, India, and Central and South America, even North America.

What is the truth about man-eating cats? First, what is a "great" cat? We normally refer to lions, tigers, leopards, jaguars, and occasionally cheetahs and mountain lions as the great cats. The cheetah, once found in Africa and Asia but now only in Africa, is never accused of being a man-eater; and the mountain lion of North, Central, and South America has been involved in only a very few such incidents in all of recorded history. Cases reported for the puma, as the mountain lion is called, almost always involved young cats, barely more than cubs. Some weighed as little as twenty-five pounds and were less than one-fifth grown. Apparently they had lost their mothers, didn't know how to hunt well enough to feed themselves, and just blundered into, and jumped on, anything that moved. They were usually killed by the person they attacked. No one can seriously think of the mountain lion as a man-eater. That is not always true of large cats.

Large cats are incredibly powerful, and man is a puny weakling in comparison. The Siberian tiger is the largest cat

on Earth. It can weigh as much as 800 pounds. If it attacked a man who did not have a high-caliber gun at the ready, a cat that size would be about as formidable as a freight train. Lions can weigh between 400 and 500 pounds and are so powerful an unarmed man could not defend himself against them. Leopards are much smaller, of course. They weigh between 150 and 250 pounds. They have powerful jaws, long canine stabbing teeth, long, raking claws, and they are stealthy and smart. They attack from ambush, and a man would be almost certain to go down before an assault. In addition, man is a very slow runner, has no claws, teeth, or kick that could help in a fight, does not see well at night, or hear well at any time. In short, man is easy prey. Since most of the people who live where the great cats live wild are too poor to own firearms and are usually unarmed, at least unarmed as far as a cat is concerned, why are so few cats man-eaters? Very simply, cats fear man.

Most large cats move away when man approaches. In fact, man has attacked cats with such great regularity that most face extinction in the next half century, while cats have rarely attacked man. It is one of the strange aspects of animal behavior as it affects us. "Insanity" is a term describing a human condition, but we assume there is a parallel condition in animals, and at the very least man-eaters are abnormal. Like renegade wolves, some cats have been "insane."

The Bay of Bengal lies east of India. To the north of the great bay lies Bangladesh. It was called East Pakistan until a recent revolution. On the southern edge of Bangladesh, in the Bay of Bengal, are the Sundarban Islands. They spread out over almost 7,000 square miles. Some are remote, and many are marshy. Tropical storms hit the area often, and before the great annual rains known as monsoons it is unbearably hot and muggy. There are days when it is almost impossible to breathe.

The very poor people in the Sundarbans simply try to survive nature's or man's dirty tricks. They have seen too many typhoons, too many wars, and too many conquerors. They have also seen too many man-eating tigers, for in the Sundarbans there have been more man-eating tigers than any other known area on Earth. No one has ever explained it. It just happens. The animals are regularly hunted by government troops or wealthy sportsmen who like to kill man-eaters. Their skins make better conversation as trophies than those of normal cats, and the hunters don't feel guilty about killing these beautiful animals. No one doubts that a man-eater must die. Also, tiger hunting is now forbidden, since the animal is an endangered species—unless it is a man-eater and public safety is at stake.

On the island of Sumatra, government troops recently killed a tigress. She had been feeding her cub human flesh, although game was plentiful. Before troops tracked her to her cave and blew her up with dynamite, she had killed and eaten seventeen people.

Jim Corbett, a hunter and writer, told many stories of his exploits hunting down man-eating tigers and leopards in India. His books are probably the best known on the subject of man-eaters: *Man-Eaters of Kumaon, The Man-Eating Leopard of Rudraprayag, The Temple Tiger,* and *More Man-Eaters of Kumaon.* Corbett offered many theories on why cats hunted man, but even he was never certain of any of them.

Tigers often try to attack porcupines. They make misjudgments, with the result that their noses, mouths, and paws are full of barbed quills that they cannot get out. The wounds become infected, and the cats are all but incapable of hunting. Unless they can find slow, lumbering, easy prey, they starve. Man fits that description. It was Corbett's contention that many if not most tigers that went bad were crippled and had no choice. Some man-eaters, he was sure, were just too old to tackle anything but a soft touch like man.

India and the surrounding countries have been subjected to terrible outbreaks of disease through the years. Plague, influenza, cholera, typhus, and typhoid fever have all struck, causing hundreds of thousands of deaths. Even now there is only one doctor for every 70,000 people, so one can imagine how much medical attention was available in remote villages fifty or seventy-five years ago. When disease struck, people simply died. Nothing could be done. Very often they died so fast survivors could not attend to their proper disposal. They didn't bury their dead, and there were too many to burn. At the peak of the crisis, the terrified survivors, abandoning sick and dead loved ones and their ancestral homes, moved away.

Cats are scavengers, as are wolves, bears, and most other predators. They eat any meat that is available. Corbett and others believed that was how some leopards became man-eaters. They would move into abandoned villages and feast on the dead and dying. When the epidemic burned itself out and those sources were gone, leopards were left with a taste for human flesh. They continued to hunt man until they themselves were located and killed, often years later.

Lions, strangely, also occasionally behave like renegade wolves. They are the only cats that live in groups and hunt together. The group, known as a pride, numbers from two or three to more than thirty. Every now and then, a cat separates itself from the pride and becomes a man-killer and even a man-eater.

Among the most famous man-eaters of all time were two lions in East Africa known as the Man-Eaters of Tsavo. Colonel J. H. Patterson, a railroad engineer who killed the cats, wrote a book by that name. Just before the beginning of this century, in 1898, two healthy young male lions began attacking crews building a railroad inland from the Indian Ocean. Their attacks became so regular and so terrifying that work on the railroad had to be stopped. There were times when they attacked every single night, seeking out the terrified crews, who huddled pitifully by their fires, listening to every sound in the darkness beyond their circle of light. After Colonel Patterson finally destroyed the lions, he examined them and found no worn teeth, no porcupine quills, and no old wounds. Two powerful young male cats had simply become the worst kind of juvenile delinquents. Many men died before they were stopped, and many risks were taken to accomplish that.

So, like the riddle of renegade wolves, this is a dual mystery. Why don't cats normally hunt man? And since they do not, what drives a few to deviate from normal behavior?

11 *The Mystery of Wadi Mukkateb*

In the Middle East, southeast of the Suez Canal, is the vast Sinai Peninsula. Most of it is stark and forbidding desert, although there are some lush oases where permanent homes and schools have been built.

For thousands of years nomads have wandered through the Sinai, Bedouins with their sheep and goats, merchants and smugglers with their camel trains, and armies out for blood and conquest. Situated as it is between Egypt, in Africa, and Israel and Saudi Arabia, in southwest Asia or the Middle East, the Sinai has been the scene of a great deal of history. People and their cultures have come and gone.

In the Middle Eastern world a wadi (pronounced, roughly, "hach-war-dee") is a dry valley or river bed. It is often barren, often surrounded by high cliffs.

Although the Sinai is next to Africa, near Saudi Arabia and on the Red Sea, it is not always as hot as we might think. A desert can be either bitter cold or burning hot. In some of the wadi in the Sinai, terribly cold winds blow in winter and early spring. One can freeze even while he is being burned to a crisp by the sun. If the wadi is high up in the barren hills and forms a natural funnel for the wind, it can be even worse.

Wadi Mukkateb is such a place, and when I wandered through it with my teen-aged son, Clay, one March day, it was cold indeed. We were burned and frozen simultaneously, but we soon became so involved in the mystery of the

place that we hardly noticed our discomfort. It is amazing how excitement can make you forget even freezing fingers and toes.

Wadi Mukkateb is a natural passageway through the rugged mountains, and obviously people have been using it as a migratory or commercial route since long, long before the time of Christ. The walls are high and red, and the floor of the canyon flat. Apparently there have been earthquakes in the area, for great blocks of sandstone have fallen from the cliffs and lie scattered across the old dry river. It is easy to see where whole sections of the cliff have slipped down. It doesn't seem likely that anything but an earthquake could have caused that kind of damage.

Amazingly, most of the portions of the cliff face that have fallen in or slipped down are covered with hundreds and hundreds of drawings, mostly of animals, and writing that has not yet been fully deciphered. By using field glasses or taking a long, dangerous climb, one can find portions of the cliff face still in position that are also covered with these strange animal drawings. No one knows who did them or why. Some of the writings, we believe, are prayers, and some just names. But the drawings are something else.

Wadi Mukkateb is never pleasant. There doesn't seem to be any sign of a settlement in the high pass. It was used for coming and going because it had to be, but no one spent much time there, no one, that is, except the people who climbed those cliffs longer ago than we know to write strange messages to their gods and draw their animal pictures.

It is likely that modern man will one day understand the strange writings. Scientists who specialize in ancient

languages usually succeed in deciphering them. Perhaps when the Wadi Mukkateb writings have been read, we will know *why* they were written. Perhaps not. In the meantime, we have this to understand: no matter what time of the year it was, it was probably uncomfortable. Men (we assume) scrambled up the cliff face and clung to the rock in what must have been very precarious positions. Surely some must have fallen and been killed or seriously injured. Still, they moved along, chipping into the rock.

It is thought that it was all done over a long period of time. Groups moving through after each other or one group that came through the area again and again stopped and contributed a share. The pictures and cartoonlike picture sequences seem to describe adventures in faraway lands and on endless travels. That would indicate that they were merchants, coming and going endlessly, and witnessing the unfolding of human history.

Perhaps that is what happened. For the moment, Wadi Mukkateb alone has the answer. The only sounds one can hear are the burning and freezing winds moving into a great funnel in the mountains and brushing past mile after mile of red sandstone before beginning the plunge down the slopes onto the flatter deserts below. One does not hear many answers in the wind—only the taunting reminder of how little we really know about our planet and its animal life and how the men and women who have observed the wonders of its nature for so many centuries still have only questions for most of its mysteries.

Epilogue: Where Does a Mystery Lead Us?

A mystery is something we do not understand, at least for the moment. We must always remember, though, that it is tied to the time in which we live. The shape of the Earth was a mystery in a different age. The number of planets in our solar system has been known only for a short period of human history. We have learned only comparatively recently that whales are not fish, that disease is caused by bacteria and viruses and not demons, and that no one lives on the moon. A thousand things one accepts and has never wondered about were profound mysteries for the most brilliant scientists only a century ago.

That means, of course, that many of the things we do not yet know will be common knowledge for our grandchildren. We double our technology and knowledge every seven years. It once took about 10,000 years to do that. Not only are we learning more and more, but we are learning it faster and faster. My son studied subjects in high school that I didn't study until I attended college.

The most exciting thing about a mystery is that it is never solved in isolation. It literally never ends, for a mystery produces new mysteries. It is a gateway into new worlds beyond.

That is why I assured you at the beginning of this book that you need never fear, as I once did, that you would grow up to face a world without mystery. There will always be mysteries. What is fascinating is that we do not know enough now even to guess what future mysteries will absorb mankind. Upcoming generations will ponder things we do not even know enough about to be curious over today. After all, it wasn't until we discovered that the world was a ball that we could begin to wonder how much it measured around the middle, what lived at the bottom and top, and how people stayed aboard on the underside.

So welcome to our world of mysteries. Some people call it nature. Some call it science. I call it one of the great joys of being alive.

Index

Alpha wolf, 45, 46, 47
Amirthamayagam, Thambi, 41, 42
Anadromous fish, 11
Antarctic, 28, 30
Arctic, 28
Atlantic salmon, 13, 14

Bangladesh, 52
Bear
  grizzly, 28
  polar, 28
Beluga, 17
Beta wolf, 45
Blue whale, 17, 18
Boar, wild, 23
Buffalo, water, 21

Canada, 35
Cape of Good Hope, 33
Caribou, 28
Ceylonese eagle-owl (devil bird), 23
Cheetah, 50
Cobra
  fangs of, 37
  hood of, 36, 37
  king, 37
  spitting, 37–38, 40
Computers, 15
Corbett, Jim, 53, 54
Cougar (mountain lion), 21, 50

Custer Wolf, 48

Devil bird (Ceylonese eagle-owl), 23
Dolphin, 17

Eagle-owl, Ceylonese (devil bird), 23
East Africa, man-eating lions in, 55
Elephant in musth, 24–27

Gila monster, 43, 44
Gravity, and salmon's migration, 15
Grizzly bear, 28

Helodermatidae (venomous lizards),
  41, 42, 43, 44

India, 53, 54

Jaguar, 50

Killer whale, 17, 19, 30
King cobra, 37
Komodo dragon, 41
Krill, 30
Kruger National Park (South Africa),
  26
Leopard, 23, 47, 50, 52
  man-eating, 53, 54
Lion, 50, 52
  man-eating, 55

Lizard(s), 41–44
  legless, 41
  meat-eating, 41
  number of species of, 41, 42
  venomous (Helodermatidae), 41, 42, 43, 44
  See also Gila monster; Komodo dragon; Mexican beaded lizard
Loch Ness monster, 35
Loris, 23

Magnetic field, Earth's, 15
Malaya, 37
Man-Eaters of Kumaon (Corbett), 53
Man-Eaters of Tsavo (Patterson), 55
Man-eating cats, 52, 53, 54, 55
Man-Eating Leopard of Rudraprayag (Corbett), 53
Mexican beaded lizard, 43, 44
Middle East, 56
Monsoon, 52
Moose, 28, 47
More Man-Eaters of Kumaon (Corbett), 53
Mosquito, wingless, in Antarctic, 28
Mountain lion, 21, 50
Musth, elephant in, 24–27
Musth gland, 27
Mystery
  life made exciting by, 7, 60–61
  never-ending, 7, 60

North Pole, 28

Old Doc (wolf), 48
Old Put's Wolf, 48
Owl-eagle, Ceylonese (devil bird), 23

Pacific salmon, 11–15
  as anadromous fish, 11
  migrations of, 11, 13, 14–15
  spawning by, 13
Panther (mountain lion), 21, 50
Patterson, J. H., 55
Penguin, 30
Polar bear, 28
Porpoise, 17, 19
Puma (mountain lion), 21, 50
Putnam, Israel, 48
Pyntz, Albany, 33

Rattlesnake, 28–39
Reindeer, 28
Renegade wolves, 48, 49
Roman Rock Island, 34, 35
Russia, 35

Salmon
  Atlantic, 13, 14
  Pacific. *See* Pacific salmon
Scotland, and Loch Ness monster, 35
Screams in nighttime forest, legends of, 20–23
Sea serpent, sightings of, 33–35
Seal, 30
  Weddell, 30, 31
Shrew, 44
Siberian tiger, 50, 52
Sinai Peninsula, 56
Skua, 30
Sloth bear, 23
Smyth, Leicester, 34
Snake(s), 36–40
  number of species of, 36
  venomous, 36, 37, 38, 39, 40, 44
  See also Cobra; Rattlesnake
South Africa
  Kruger National Park in, 26

South Africa *(cont.)*
   sea serpents sighted along coasts of, 34–35
South Pole, 28, 30, 31
Sperm whale, 17
Spitting cobra, 37–38, 40
Sri Lanka (Ceylon), 21, 24, 41
Stranded whales, 19
Sumatra, 53
Sundarban Islands, 52, 53
Sweden, 35

Temple Tiger, The (Corbett), 53
Tern, 30
Thailand, 37
Three Toes (wolf), 48
Tiger, 47, 50
   man-eating, 53
   porcupine attacked by, 53
   Siberian, 50, 52
Typhoon, 53

Wadi Mukkateb, 56–57
   drawings and writings on cliff face of, 57–59
Water buffalo, 21
Weddell seal, 30, 31
Whale(s)
   blue, 17, 18
   intelligence of, 17
   killer, 17, 19, 30
   size of, 18
   sonar signals of, 17, 18
   sperm, 17
   stranded, 19
Wolf, 45–49
   Alpha, 45, 46, 47
   Beta, 45
   fear of, unwarranted, 48–49
   as hunter, 47
   as parent, 47
   renegade, 48, 49
   as social animal, 45
World of Wonders, A (Pyntz, ed.), 33